Jeff Conquers
the Cube!

Jeff Conquers the Cube!

in 45 Seconds*

*And You Can, Too

Jeffrey Varasano

day books

A Division of STEIN AND DAY/Publishers/New York

For Arlene DeSimone and Jim Cohen,
teachers at Truman High School

Third printing 1981
First published in 1981
Copyright © 1981 by Jeffrey Varasano
All rights reserved
Printed in the U.S.A.

STEIN AND DAY/*Publishers*
Scarborough House
Briarcliff Manor, N.Y. 10510
ISBN 0-8128 7097-2
Library of Congress Catalog Card Number 81-480-23

Acknowledgments

I would like to give special thanks to my uncle, Robert R. Reichin, for his enthusiasm and support.

Introduction

Congratulations! You've taken the first step in solving your Rubik's ™ Cube. My name is Jeffrey Varasano, and I am 15 years old. I first played with a Rubik's™ Cube a year ago. After two months, I had it conquered. I still practice often, and I now have an average solution time of 40 seconds. My record time is 25 seconds.

Several of my friends, after reading solutions in other publications, came to me and asked me if I could write a solution that was shorter and less complicated than the others.

Here is my answer to their requests:

JEFF CONQUERS THE CUBE.

Throughout these instructions, you will find that letters play a major role in your journey toward solution. Letters are used both to *label* sections of the cube and to serve as *guides for moves*.

DIAGRAM 1A

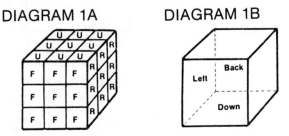

DIAGRAM 1B

The letters stand for: R (right), L (left), F (front), B (back), U (up), and D (down). This system is illustrated in Diagrams 1A and 1B.

To further simplify this introduction, I will illustrate other portions of the cube and certain terms which will be used throughout these instructions.

DIAGRAM 2

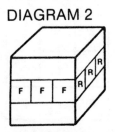

DIAGRAM 3

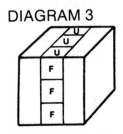

In DIAGRAM 2, the HORIZONTAL slice is illustrated.

The VERTICAL slice

The UP slice

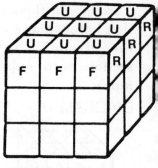

DIAGRAM 4

The RIGHT slice

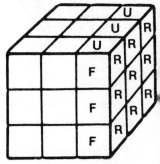

DIAGRAM 5

The FRONT slice

DIAGRAM 6

The LEFT slice

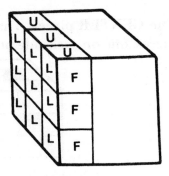

DIAGRAM 7

The DOWN slice

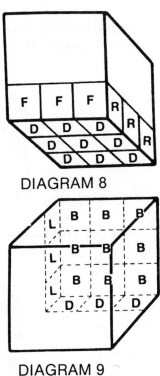

DIAGRAM 8

The BACK slice

DIAGRAM 9

11

The CENTER pieces. There are six of these pieces, one on each side of the cube.

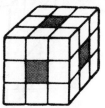

DIAGRAM 10

The MIDDLE-EDGE pieces. There are 12 of these pieces on the cube.

DIAGRAM 11

The CORNER pieces. There are eight of these pieces on the cube.

DIAGRAM 12

NOTE: Any piece can be referred to by the letters of the sides it is located on. For example, the UR piece is so named because it is located on the U (upper) and R (right) sides.

In DIAGRAM 13, the UR piece is darkened.

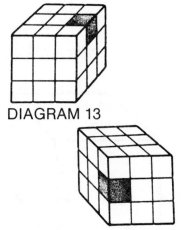

DIAGRAM 13

The FL piece

DIAGRAM 14

NOTE: A CORNER piece is also described in terms of the sides on which it is located.

The FUR corner piece

DIAGRAM 15

The BU piece

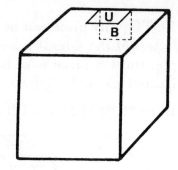

DIAGRAM 16

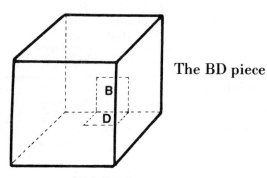

The BD piece

DIAGRAM 17

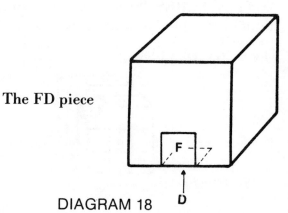

The FD piece

DIAGRAM 18

14

I have used a simple system for guiding you through each move:

> The RIGHT SLICE (see Diagram 5) will be indicated by an R.
>
> The LEFT SLICE (see Diagram 7) will be indicated by an L.
>
> The FRONT SLICE (see Diagram 6) will be indicated by an F.
>
> The DOWN SLICE (see Diagram 8) will be indicated by a D.
>
> The BACK SLICE (see Diagram 9) will be indicated by a B.
>
> The UP SLICE (see Diagram 4) will be indicated by a U.
>
> The HORIZONTAL SLICE (see Diagram 2) will be indicated by a horizontal arrow.
>
> The VERTICAL SLICE (see Diagram 3) will be indicated by a vertical arrow.

Directions of moves:

If the letter is indicated alone (i.e., R), then simply turn the side that the letter stands for *clockwise*. (Imagine that there is the face of a clock on the side you are turning. Turn the side in the direction that the hands of the clock would move.)

If a letter has an apostrophe following it (i.e., R'), turn the side that the letter stands for *counterclockwise*.

See the diagrams below for examples.

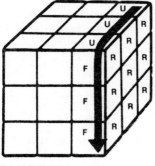

DIAGRAM 19

R' (Right side counterclockwise)

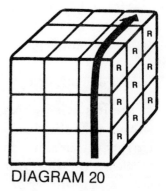

R (Right side clockwise)

DIAGRAM 20

If a letter has a number 2 following it (i.e., R^2), then turn the side that the letter stands for twice. When turning a side twice, it makes no difference whether you turn it clockwise or counterclockwise.

Below are diagrams showing clockwise moves for the UP, DOWN, RIGHT, and FRONT slices. You can refer to these diagrams later on if you get confused.

DIAGRAM 21

FRONT slice clockwise turn

DIAGRAM 22

UP slice clockwise turn

DIAGRAM 23

RIGHT slice clockwise turn

DIAGRAM 24

DOWN slice clockwise turn

In addition, if you see a **HORIZONTAL ARROW** (i.e., ← or →), turn the **HORIZONTAL** slice (see Diagram 2) in the direction that the arrow is pointing.

If you see a **VERTICAL ARROW** (i.e., ↑ or ↓), turn the **VERTICAL** slice (see Diagram 3) in the direction that the arrow is pointing.

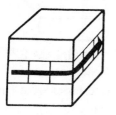

→ DIAGRAM 25

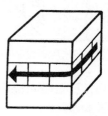

← DIAGRAM 26

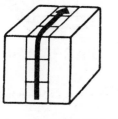

↑ DIAGRAM 27

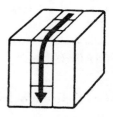

↓ DIAGRAM 28

It should be noted that different cubes have different arrangements of colors. When you buy your Rubik's cube, the red side may be opposite the orange, or perhaps the blue side is opposite the orange, or otherwise not the same as another cube. Therefore, instead of referring to specific colors, I am simply going to refer to Color 1, Color 2, Color 3, etc. Start with whatever color you want, and call it Color 1.

You are now ready to take the first step leading to the solution.

STEP I

The first approach is to prepare one "X" of Color 1 (Diagram 29).

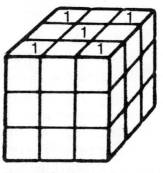

DIAGRAM 29: An "X"

Turn your cube so that the Color 1 CENTER piece is on the UP slice (Diagram 30).

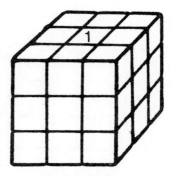

DIAGRAM 30

If any corners of the X are already in place, make sure they are not in the FUR (Front-Upper-Right) position (Diagram 31).

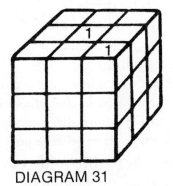

DIAGRAM 31

If there *is* a Color 1 piece on top, and in the FUR position, turn the U slice (the Upper Slice) until the slot has a different color in it (Diagram 32). We now say that the slot is "empty."

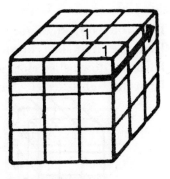

DIAGRAM 32

STEP IA

If there is a color 1 piece in position 1 (Diagram 33), make the following move:

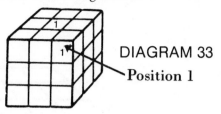

DIAGRAM 33
Position 1

Move for position 1:
1. Turn the right slice clockwise (R)
2. Turn the front slice clockwise (F)
3. Turn the down slice counterclockwise (D')
4. Turn the front slice counterclockwise (F')

Now turn the U slice so that the FUR slot is empty again. Now go back to Step A, to see if you have a new Color 1 piece in position 1. If not, move to B.

STEP IB

If there is a Color 1 piece in position 2 (Diagram 34), make the following move (if there is no Color 1 piece in position 2, go to C.):
1. Turn the front slice counterclockwise (F')
2. Turn the right slice counterclockwise (R')
3. Turn the down slice clockwise (D)
4. Turn the right slice clockwise (R)

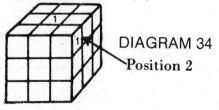

DIAGRAM 34
Position 2

22

Turn the U slice so that the FUR slot is empty again. Now go back to A to see if you have a new Color 1 piece in position 1. If not, move to B.

STEP 1C

If there is no Color 1 piece in positions 1 or 2, then there must be a Color 1 corner piece somewhere on the DOWN slice. Turn the DOWN slice so that a Color 1 piece is in one of the FDR positions. Now the piece will be either in position 3 (Diagram 35), position 4 (Diagram 36), or position 5 (Diagram 37).

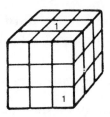

Position 3

DIAGRAM 35

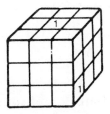

Position 4

DIAGRAM 36

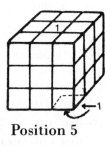

Position 5

DIAGRAM 37

If the piece is in position 3, then: *front*

1. Turn the front slice clockwise (F)
2. Turn the down slice clockwise (D)
3. Turn the front slice counterclockwise (F')

Now turn the U slice until the FUR slot is empty again, and return to A.

If the piece is in position 4, then: *side*

1. Turn the right slice counterclockwise (R')
2. Turn the down slice counterclockwise (D')
3. Turn the right slice clockwise (R)

Now turn the U slice until the FUR slot is empty again, and return to A.

If the piece is in position 5, then: *bottom*

1. Turn the down slice counterclockwise (D')
2. Turn the right slice twice (R^2)
3. Turn the down slice clockwise (D)
4. Turn the right side twice (R^2)

Now turn the U slice until the FUR slot is empty again, and return to A.

After you have filled in all four slots of your X, move on to Step II. (Several middle-edge Color 1 pieces might also be in place. Just leave them and go on to Step II.)

STEP IIA

Place the X of Color 1 face down, so it now becomes the DOWN slice. Whatever color is in the new UP slice center piece will determine the color of that slice, and this is the color that you will be matching in the following diagrams. We call this Color 2.

One of the following patterns of corner pieces (Diagrams 38-44) will appear on your UP slice. Don't worry about the colors of any pieces but the corner pieces.

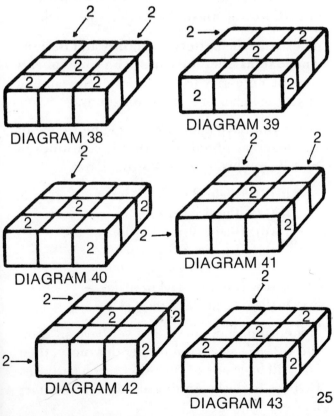

DIAGRAM 38

DIAGRAM 39

DIAGRAM 40

DIAGRAM 41

DIAGRAM 42

DIAGRAM 43

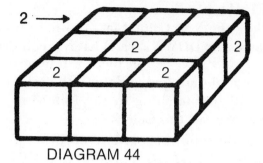

DIAGRAM 44

IMPORTANT—If you find that you have a pattern that matches diagram 44, substitute $R^2F^2R^2$ for the move that is described below.

Look over the patterns carefully and find the one that matches the pattern on your cube. It may be necessary to turn your entire cube so that it matches one of the diagrams.

After you have matched your cube to one of the diagrams, hold it so that it continues to match, and make the following move:

1. (R)   (Right slice clockwise)
2. (U^2)   (Up slice twice)
3. (R)   (Right slice clockwise)
4. (U^2)   (Up slice twice)
5. (R)   (Right slice clockwise)
6. (U)   (Up slice clockwise)
7. (R^2)   (Right slice twice)

If you find that you do not yet have an X on the UP slice after completing this move, then you will have one of the other seven patterns. Go back to the beginning of this step and start over. You may have to go through Step IIA three times. When you find that you have an X on the UP slice, then you are finished with this step. Go to Step III.

STEP IIB—For the Expert:

This step serves the same purpose as Step IIA. The only difference is that it is faster. Instead of repeating the same move over and over, you need perform only one move. The use of this alternate method has one advantage and one drawback. By using this method, you will decrease the number of moves needed to reach the solution. On the other hand, you will need to memorize more moves.

Place your X face down. The Color 1 is on the D slice. On the top slice, one of the following seven patterns will appear. The Center piece of the top slice determines the color of that slice and this is the color that you will be matching to the following diagrams. Look over the patterns carefully and find the one that matches the pattern on your cube. It may be necessary to turn your entire cube so that it matches one of the diagrams. After you have matched your cube to one of the diagrams, do the move to the right of it.

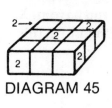

DIAGRAM 45

1. R
2. U²
3. R'
4. U'
5. R
6. U'
7. R'

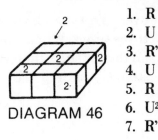

DIAGRAM 46

1. R
2. U
3. R'
4. U
5. R
6. U²
7. R'

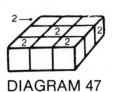

DIAGRAM 47

1. R
2. U
3. R'
4. U'
5. F'
6. U'
7. F

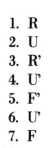
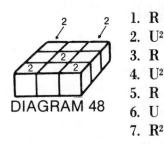

DIAGRAM 48

1. R
2. U²
3. R
4. U²
5. R
6. U
7. R²

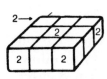

DIAGRAM 49

1. R
2. U
3. R²
4. F'
5. R²
6. U
7. R'

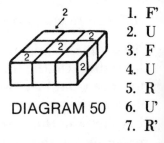

DIAGRAM 50

1. F'
2. U
3. F
4. U
5. R
6. U'
7. R'

After making your move, another X will appear on the UP slice, while the X on the DOWN slice will remain intact. Now go to Step III.

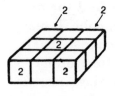

DIAGRAM 51

1. R²
2. U²
3. R
4. U²
5. R²

28

STEP III

You now have an X on the top and bottom of your cube. In this step you will work with the remaining four sides of your cube to match the colors of the corner pieces. When two corner pieces are matched, you have what I call a "correct edge" (Diagrams 52–53).

Correct edge ⟶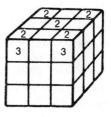

DIAGRAM 52

Incorrect edge ⟶

DIAGRAM 53

When you are done, there will be two pairs of correct edges on each side of your cube, one pair on the top, and one on the bottom (Diagram 54).

DIAGRAM 54

All four sides will look like this.

STEP IIIA

Count the total number of correct edges that you already have on your cube.

1. If you have no correct edges, make the following move:
    1. R^2
    2. F^2
    3. R^2

    You will now have all your correct edges. Proceed to Step IV.

2. If you have one correct edge, make the following move:
    1. R^2
    2. F^2
    3. R^2

    You will now have five correct sides. Proceed to Step IIIA (5).

3. If you have two correct edges, turn the U slice until the one correct edge on the U slice is on the front side. Now repeat the following move five times:
    1. R^2
    2. F^2
    3. U

    You will now have five correct edges. Proceed to Step IIIA (5).

4. If you have four correct edges, hold the cube so that the four correct edges are on the Down slice. (It may be necessary to turn

the cube over.) Now repeat the following move five times:

1. R²
2. F²
3. U

You will now have five correct edges. Proceed to Step IIIA (5).

5. If you have a total of five correct edges, hold the cube so that four correct edges are on the Down slice and the one correct edge on the Up slice is on the front side. (It may be necessary to turn your cube over in order to hold the cube properly.) Now repeat the following move five times:

1. R²   3. U
2. F²

You now have your correct edges. Proceed to Step IV.

6. If you have eight correct edges, you should proceed directly to Step IV.

STEP IIIB—For the Expert

This step serves the same purpose as Step IIIA. The only difference is that instead of repeating the same move over and over, you need to perform only one move. The use of this alternate method has one advantage and one drawback: You will decrease the number of moves needed to reach the solution, but you will need to memorize more moves.

Count the total number of correct edges on each X. You now have no, one, two, four, five, or eight correct edges. It may be necessary to turn the cube in order to hold it the way you are directed. You may also have to turn the cube over.

1. If you have eight correct edges, go directly to Step IV.

2. If you have five correct edges, hold the cube so that you have four correct edges on the DOWN slice and so that the one correct edge on the U slice is on the left slice. Now do the following move:

1. R^2	5. R^2	8. B^2
2. U'	6. B^2	9. U'
3. R^2	7. U	10. B^2
4. U		

You now have your correct edges. Proceed to Step IV.

3. If you have four correct edges, then hold the cube so that the four correct edges are on the D slice X. Now do the following move:

1. B^2	5. R^2	9. F^2
2. U^2	6. U^2	10. U^2
3. R^2	7. F^2	11. L^2
4. U	8. U	

You now have your correct edges. Proceed to Step IV.

4. If you have two correct edges, then turn the U slice and/or the D slice until each correct

edge is on the front slice. Now do the following move.

1. R^2 5. F^2
2. U   6. U
3. R^2 7. F^2
4. U^2

5. If you have one correct edge, then hold the cube so that the one correct edge is on the D slice, and turn the cube so that the correct edge is on the left slice. Now do the following move:

1. R^2 4. U   7. R^2
2. U'   5. R^2 8. U
3. R^2 6. U'   9. R^2

You now have your correct edges. Proceed to Step IV.

6. If you have no correct edges, then make the following move:

1. R^2
2. F^2
3. R^2

You now have your correct edges. Proceed to Step IV.

STEP IV

You will now be filling in the MIDDLE EDGE PIECES of the UP and DOWN slices (see Diagram 11). The correct middle edge piece matches the color of the corner pieces and the center piece of the UP slice, but not necessarily to the center piece of the RIGHT slice (Diagram 55).

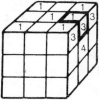

DIAGRAM 55
Matching
middle
edge piece

Throughout this step you will be working with the UR position (Diagram 56) to fill it in with the proper colors.

UR

DIAGRAM 56

STEP IVA

Hold your cube so that the Color 1 X is on the UP slice. Find the UR position. You will be looking for the piece that will fit (match) in the UR position. In Diagram 57 that would be the piece with colors 1 and 3 (the 1, 3 piece).

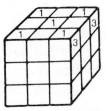

DIAGRAM 57

34

Is the piece that belongs in the UR position somewhere on the HORIZONTAL slice? (See Diagram 2.) If the 1, 3 piece is on the HORIZONTAL slice, then turn that slice until the piece is in the FL position (Diagrams 58 and 59).

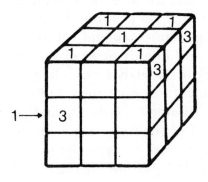

DIAGRAM 58

Or, the 1, 3 piece may be like this:

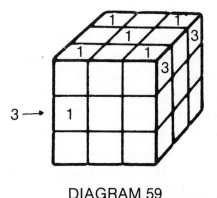

DIAGRAM 59

side

1. If the piece is the same as shown in Diagram 58, make the following move:
   1. R'
   *2. →
   3. R

   Now turn the UP slice until an empty slot appears in the UR position. Return to the beginning of this step and fill in the new UR slot. (Fill in only *three* of the four UP slice MIDDLE EDGE slots.) When you finish filling in three slots, turn the cube over so that the Color 2 X is on top and the one empty slot of the Color 1 X is in the DR position. Be sure to keep the one empty slot in the DR position for the rest of this step. Next you return to the beginning of this step and fill in *all four* of Color 2's MIDDLE EDGE slots.

2. If the piece is the same as shown in Diagram _front_ 59, then make the following move:
   1. R
   *2. →²
   3. R'

   Now turn the UP slice until an empty slot appears in the UR position. Return to the beginning of this step and fill in the new UR slot. (Fill in only *three* of the four slots. See Step IVA (1) above.)

---

* Remember that this arrow means you turn the horizontal slice to the right.

STEP IVB

Is the piece that belongs in the UR position somewhere on the DOWN slice? If it is, turn the DOWN slice until the piece is in the DR position. It will look like either Diagram 60 or 61.

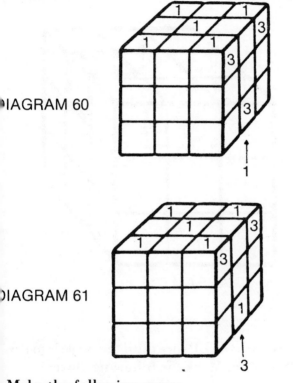

DIAGRAM 60

DIAGRAM 61

Make the following move:

1. R
2. ←
3. R'

Now return to Step IVA.

STEP IVC

Is the piece that belongs in the UR position somewhere on the UP slice, in one of the *wrong* UP slots? (Diagram 62.)

Incorrect slots   Correct slo

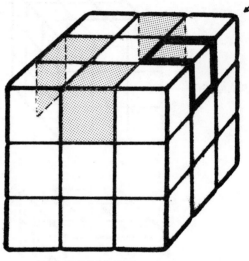

DIAGRAM 62

If it is, turn the UP slice until the piece is in the UR position. Make the following move:

1. R'
2. ←
3. R

Now turn the U slice so that the correct slot is again in the UR position. Go back to Step IVA.

STEP IVD

Is the piece that belongs in the UR position already in the UR slot, but flipped* from its correct position? (Diagram 63.)

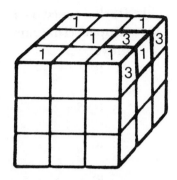

DIAGRAM 63

Flipped. A piece is flipped when it is in the correct slot, but is in backward. For example, in the above diagram the 1, 3 piece is in the UR position, but the color 3 is on the U side and the color 1 is on the R side. This is the reverse of what it should be.

If the piece if flipped, make the following move:

1. R'
2. ←
3. R

Now go back to Step IVA.

STEP IVE

Is the slot already filled? If it is, turn the top until there is an empty slot in the UR position and go back to Step IVA.

After you have filled in all four MIDDLE EDGE slots for Color 2, you are ready to complete the final piece for Color 1. Turn the cube back over and proceed as follows:

1. If the piece is in the UR position, but flipped (Diagram 64), make the following move:

   1. R'
   2. ←
   3. R
   4. ←
   5. R
   6. ←
   7. R'

   Now go to Step V.

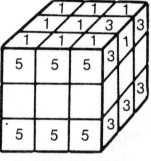

DIAGRAM 64

2. If the final piece is on the HORIZONTAL slice, turn that slice until the piece is in the FL position (Diagrams 65–66).

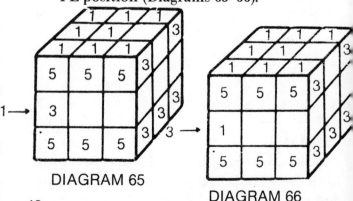

DIAGRAM 65

DIAGRAM 66

If the piece appears as shown in Diagram 65,
make the following move:

1. ←
2. R
3. →
4. R²
5. →
6. R

If the piece appears as shown in Diagram 66,
make the following move:

1. R'
2. ←
3. R²
4. ←
5. R'

Now go to Step V.

STEP IVF—For the Expert

It is easiest and fastest to fill a piece into the UR slot if it is already on the HORIZONTAL slice (see Diagrams 58 and 59). Look at Diagram 66A. Normally, you would look for the 1, 3 piece and go through several moves in order to put it in the UR slot. However, if you simply turned the U slice once, counterclockwise, you would now have Diagram 66B.

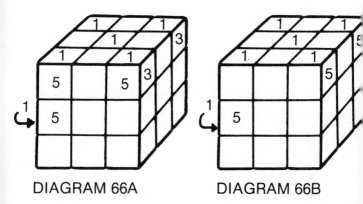

DIAGRAM 66A   DIAGRAM 66B

Then you can simply follow the moves for a UR piece on the HORIZONTAL slice (Step IVA). This will shorten your moves and speed up your completion time.

STEP V

Now turn the U, D, and HORIZONTAL slices so that the edges and center pieces of the same color are on the same side (Diagram 67).

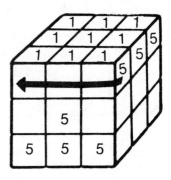

DIAGRAM 67

Turn the cube over so that the two complete sides are now the Right and Left sides. Find the piece that belongs in the FU position (Diagram 68).

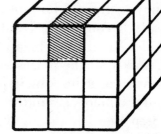

DIAGRAM 68
FU position

43

STEP VA

If the FU piece is already in the proper position—even if it is flipped—go to VE.

STEP VB

If the piece is in the FD position (See Diagram 18), do the following move:

1. ↓   (Remember, this symbol directs
2. D^2 you to turn the Vertical slice in
3. ↑   the indicated direction)
4. D^2

Now go to Step VE.

STEP VC

If the piece is in the BU position (see Diagram 16), do the following move:

1. ↑   3. ↓
2. B^2 4. B^2

Now go to Step VE.

STEP VD

If the piece is in the BD position (see Diagram 17), do the following move:

Now go to Step VE.

1. B^2 3. B^2
2. ↑   4. ↓

STEP VE

After making the correct move, the proper piece should be in the FU slot. It may or may not be flipped. Roll the cube over so that the U side becomes the F side. The two complete sides should still be on the right and the left. Now, if all the pieces are not yet in the right position, go back to Step VA one more time. If all the pieces *are* in the correct position, go to Step VF.

44

TEP VF

You will now have no, two, or four pieces
ipped. If you have two pieces flipped and
hey are both on the same side, hold the cube
o that the pieces are in the UR and UL posi-
ions. Now do the following FLIPPING
move:

1.	R'	7.	R
2.	←	8.	→²
3.	R²	9.	R²
4.	→²	10.	→
5.	R'	11.	R
6.	U²	12.	U²

' you have two pieces flipped and they are
iagonal from each other, hold the cube so
hat the two flipped pieces are in the UL and
he DR positions. Turn the Right slice twice
R²). Perform the FLIPPING move described
bove. At the end of the move, do R² again.

f all four pieces are flipped, hold the cube so
hat the two complete sides are the FRONT
nd BACK sides. Now do the FLIPPING
move. Turn the cube over so that the two
lipped pieces are in the UL and UR positions
nd repeat the FLIPPING move.

CONGRATULATIONS! YOU HAVE NOW
CONQUERED THE CUBE!

Appendix

TEST YOUR SKILLS

Once you've mastered Rubik's Cube, you may
want to challenge yourself by trying to make
the following patterns. (Each unseen side
should be the mirror image of the diagram.)

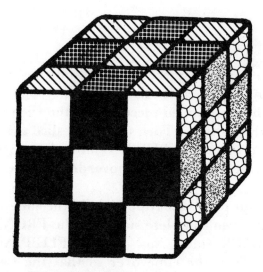

Checkerboard Pattern

Hint: Turn each center slice twice.

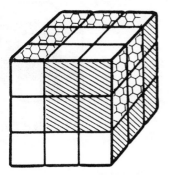

The 3-Color Cube
Within the Cube

The Lonesome Corner

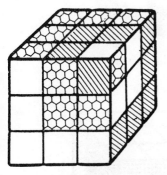

llow the Yellow Brick Road

Dots

The
ree Stooges

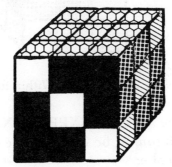

47

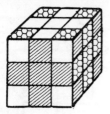

6-Sided Cross

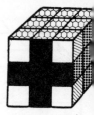

4-Sided Cross

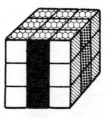

Parallel Lines

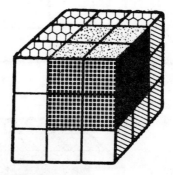

The 6-Color Cube Within the Cube

For the SUPER EXPERT! Note that this pattern *cannot* be made on the opposite side of the cube.